Mathaus Henrique da Silva Alves

Analysing the Characteristic Curves of Ltspice Solar Cells

Mathaus Henrique da Silva Alves

Analysing the Characteristic Curves of Ltspice Solar Cells

Solar Simulation

ScienciaScripts

Cover image: www.ingimage.com

This book is a translation from the original published under ISBN 978-613-9-70189-6.

Publisher:
Sciencia Scripts
is a trademark of
Dodo Books Indian Ocean Ltd. and OmniScriptum S.R.L publishing group

120 High Road, East Finchley, London, N2 9ED, United Kingdom
Str. Armeneasca 28/1, office 1, Chisinau MD-2012, Republic of Moldova, Europe
Printed at: see last page
ISBN: 978-620-8-16654-0

ACKNOWLEDGEMENTS

To my beloved mum, Maria Sueli Simões da Silva, for her teachings, thanks to which I remain firm and thirsty to learn, and thus continue my studies.

To my advisor Prof Dr Thiago Franchi Pereira da Silva, for agreeing to advise me, for proposing this topic and for his support during the difficulties, undoubtedly a great professional.

To the professors at the UFVJM Janaúba campus, for their quality teaching, which has contributed to my knowledge and world view.

To UFVJM, PROEXC and FAPEMIG, for having provided my professional and personal growth, which contributed to the development of this work.

To my family and friends, for helping me with my studies, even in the midst of difficulties.

To Professors Dr Thiago de Lima Prado and Dr Ananias Borges Alencar for their suggestions and corrections.

SUMMARY

In describing the operation of photovoltaic devices, equivalent circuit modelling processes are widely used, in which mathematical equations describe the electrical parameters of the panels.

To find the parameters associated with photovoltaic panels, it is necessary to solve equations with a large number of variables, which makes it essential to use computer resources to solve numerically and graph the results.

Computer simulation of solar cells is therefore necessary in order to describe and improve the photovoltaic system, i.e. to determine the best parameters and operating conditions in order to create more efficient cells.

Solar simulation is useful for various purposes, such as analysing and studying the behaviour of the power converters attached to the system, simulating the behaviour of the maximum power point tracker and estimating the efficiency of the photovoltaic system.

In addition, the simulation can also be used for various operating conditions, such as partial shading, changing irradiance and fault conditions.

This work therefore aims to study the behaviour of conventional silicon photovoltaic cells using Ltspice simulation.

This paper presents two equivalent circuit models, which were used to evaluate the effects of electrical parameters on the characteristic curves of the photovoltaic panel used by GNOATTO, et al. (2005), who determined them experimentally under field conditions.

The simulation was satisfactory as the relative error was low compared to the experimentaldata in the article, around 3.45 % for the first simulation and negligible for the second simulation. This model can be adapted to characterise dye-sensitised and organic cells, which are still under development and are promising for the current market.

Key words: Simulation. Solar cells. Computer modelling. Equivalent circuit.

SUMMARY

CHAPTER 1

INTRODUCTION

The use of solar energy has advanced steadily due to concerns about energy depletion and environmental problems caused by the extraction and use of fossil fuels, where it has become a viable energy alternative for systems connected to electricity grids.

In this context, the construction of more efficient solarcells has been the subject of research worldwide, leading to the use of different technologies and alternative material resources, prioritising the best conversion of solar energy into electricity [1].

In 2001, photovoltaic energy in Brazil was limited in its use, as it was expensive and restricted the consumption of users, being used more in isolated locations [2].

However, there are now more than 27,000 solar photovoltaic systems connected to the electricity grid producing 252 MW of power, which represents more than R$1.9 billion in accumulated investments since 2012.

This progress is the result of the inauguration of photovoltaic plants and individual consumers, where 77.4 per cent of the units installed are in homes, as this type of technology has become more affordable [3].

Another relevant point, apart from the cost-benefit ratio, is that in 2012 the National Electric Energy Agency (ANEEL) regulated a system called the electricity compensation system, which makes it possible to pass on the surplus energy produced to the public energy distribution network in exchange for a discount on the electricity bill, which has also led to the growing use of photovoltaic energy in homes [4].

Thus, with the growth of the photovoltaic industry and the increase in the number of panels installed around the world, there has been a need to simulate photovoltaic panels under various operating and climatic conditions.

Photovoltaic module manufacturers only provide electrical parameters for standard test conditions (STC) with irradiation G = 1000 W / m^2 , cell temperature TC = 25 °C and air mass AM = 1.5 [5]. The standard temperature of 25 °C is rarely reached in real operating conditions, on clear sunny days it is typically 20 to 40 °C higher than ambient [2].

When describing and studying the operation of photovoltaic devices, it is common to use equivalent circuit modelling, in which mathematical equations are used to describe the electrical parameters of the panel.

To find these parameters it is necessary to solve equations with a large number of variables, which makes it essential to use computer resources to solve numerically and graph the results [6].

Simulation is therefore necessary in order to improve and create more efficient solar cells, determining the best working condition for various operating conditions, such as partial shading, changing irradiation, fault conditions, and also to advance in the construction of the most up-to-date cells, such as those sensitised by dyes and organic [5, 6].

This work aims to study the behaviour of conventional silicon solar cells using Ltspice simulation, where the effects of the electrical parameters on the characteristic curves were evaluated, and then the parameters of the I-V curves of the photovoltaic panel used by GNOATTO, et al. (2005) [2], which determined them experimentally, were extracted.

The Federal University of the Jequitinhonha and Mucuri Valleys (UFVJM) does not yet have silicon cells for research and analysis, so the work was limited to simulating the curves obtained from the author's article [2].

CHAPTER 2

THEORETICAL FRAMEWORK

2.1 PHOTOVOLTAIC CELLS

Photovoltaic cells are devices that produce electrical energy from solar radiation, a process known as the photovoltaic effect that was observed by Edmond Becquerel in 1839. The conversion takes place through the absorption of light from a structure made of a semiconductor, with an electrical potential difference appearing at the ends of the structure (Figure 1) [7].

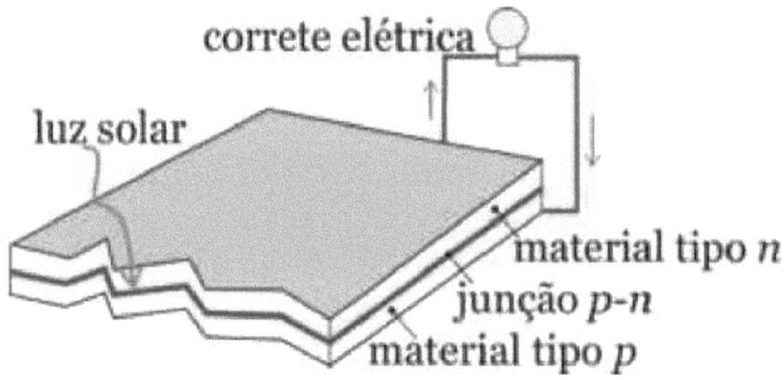

Figure 1- Working principle of photovoltaic cells

Source: [8]

Solar cells are mainly made up of semiconductor materials that have energy bands, which are the valence bands, the energy level where electrons are weakly bound, and the conduction bands, where electrons are able to conduct electricity. Photovoltaic cells need the energy of the sun's photons to excite electrons from the valence band to the conduction band, which with the combination of materials called dopants[1] generates an electrical potential difference [10, 11].

The energy required for conduction or separation of the bands is called the

[1] Doping materials are impurities added to a silicon crystal lattice, for example [9].

energy gap and is in the order of 1 eV, which differentiates semiconductors from insulators, where the gap in insulators is several eVs. Therefore, photons in the visible range, with energy higher than the material's gap, excite electrons in the conduction band. This effect can be observed in pure semiconductors, also known as intrinsic semiconductors, such as pure silicon and germanium, but they are not able to guarantee the functioning of photovoltaic cells on their own, as they only become a conductor due to the increase in temperature caused by photons. It is necessary to dope the semiconductor, forming an appropriate doped structure so that the excited electrons can generate a useful current [12].

Silicon is currently the most widely used semiconductor in the manufacture of photovoltaic cells, its atoms forming a crystalline network through bonds with atoms This structure is combined with other materials (impurities) to form the pn junction. In other words, dopants such as phosphorus and boron, which have five bonding electrons and three bonding electrons respectively, are added to this network (Figure 2). In the case of phosphorus or dopant n, there is one electron left that is weakly bound to its atom of origin, which means that, with the energy of the incident photons, this electron frees itself and moves into the conduction band. The boron or p dopant, on the other hand, needs an electron to satisfy the bonds with the silicon atoms in the lattice, so a little thermal energy leads to the conduction of a neighbouring electron that occupies the position that had a gap. This movement of electrons in the pn junction (Figure 3) generates a potential difference [10, 12].

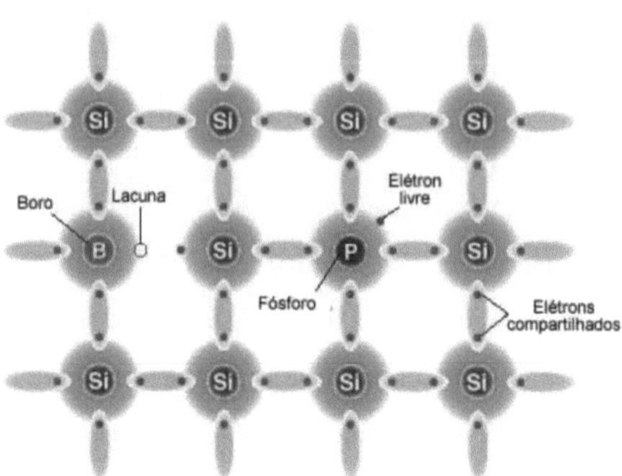

Figure 2 - Silicon doped with boron and phosphorus

Source: [13]

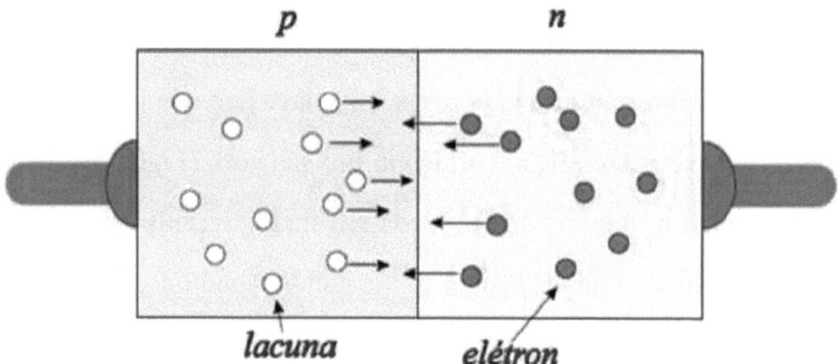

Figure 3 - Electron movement in the pn junction
Source: [14]

The materials in solar cells are chosen to utilise most of the available solar spectrum and have the lowest production cost. For this reason, silicon is a common choice because its absorption characteristics are suitable for the solar spectrum and the manufacturing technology is well developed [15]. Normally, solar cells produce low power, approximately 2 to 3 watts, requiring several cells connected in series and parallel to form modules, which connected together form panels increasing the power produced (Figure 4) [11].

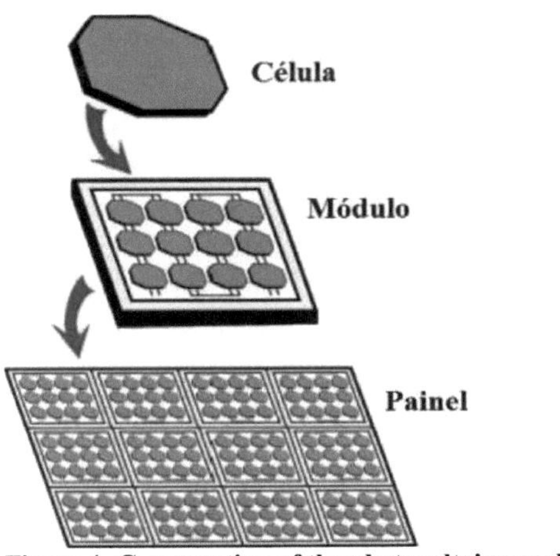

Figure 4- Construction of the photovoltaic panel

Source: [11]. Adapted.

2.2 TYPES OF PHOTOVOLTAIC CELLS

2.2.1 First Generation Cells

There are three generations of solar cells, the first generation comprising single pn junction cells made of monocrystalline or polycrystalline silicon (Figure 5). These cells account for 90 per cent of the photoconversion market. Polycrystalline silicon (p-Si) has an efficiency of between 11 and 16 per cent and is the most widely used material in the construction of photovoltaic devices on the market. Cells with monocrystalline silicon (mono-Si) have a higher efficiency of around 20%, with the current efficiency record being 25.6% [16].

The monocrystalline silicon cell is manufactured by extracting the silicon dioxide crystal, which is deoxidised, purified and solidified to a purity of 98 to 99%. As this level of purity has a perfect crystal structure, it makes production expensive and complex, and the yield of these plates is high. In the case of polycrystalline silicon cells, the manufacturing process is similar to that of monocrystalline silicon cells, but the preparation process is less rigorous, making these cells a little cheaper.

However, despite reducing the cost of production, the polycrystalline silicon cell has impurities, so there is a loss of efficiency [7, 10, 17, 18].

Silício policristalino **Silício monocristalino**

Figure 5 - Crystalline Silicon Cells

Source: [17]. Adapted.

2.2.2 Second Generation Cells

The second generation seeks to reduce production costs. Therefore, its main advantage is its manufacturing cost, which does not depend on sophisticated production techniques. However, they are generally less efficient than the first generation and have a shorter service life. Some constructions are amorphous silicon (a-Si), cadmium telluride (CdTe) and copper-indium-gallium selenide (CIGS) cells. CIGS is an exception and has a record efficiency for thin films of 20.5% yield, an efficiency close to the mono-Si cell [16, 17].

Of the silicon cells, compared to the first generation, a-Si (Figure 6) contains a high degree of disorder in the structure of the atoms, which is why it has a low production cost and low efficiency. Another relevant point is that the amorphous silicon cell, when exposed to bad weather, is affected by a degradation process, reducing its efficiency in the first few months of application. However, with their low production cost, these cells can be viable for certain applications [7, 10, 18].

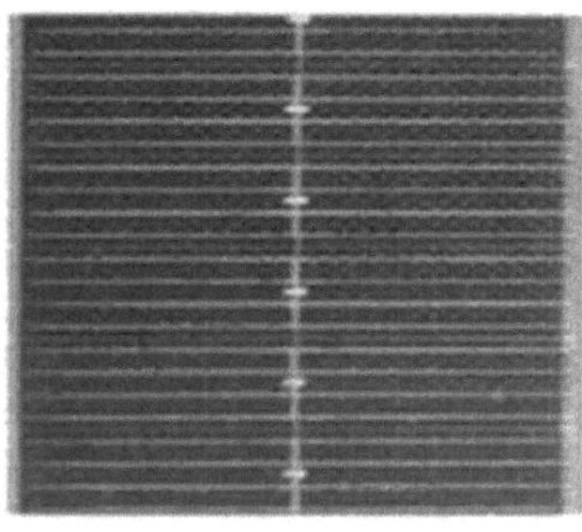

Figure 6 - Amorphous Silicon Cell (a-Si)
Source: [7]

2.2. 3Third generation cells

Third generation solar *cells* are *dye-sensitised* cells (DSCs), multi-junction cells and *organic* photovoltaics (OPVs). OPVs use polymers or small organic molecules that absorb light, DSCs use organic dyes to improve the use of the solar spectrum, while multi-junction cells try to maximise efficiency by using various semiconductor materials with different *pn* junctions. The multi-junction cell model in Figure 7 has three *pn junction* layers, where each semiconductor in the stack has a different bandgap, which in response to different wavelengths of light, aims to optimise the efficiency of absorbing solar radiation [16, 17].

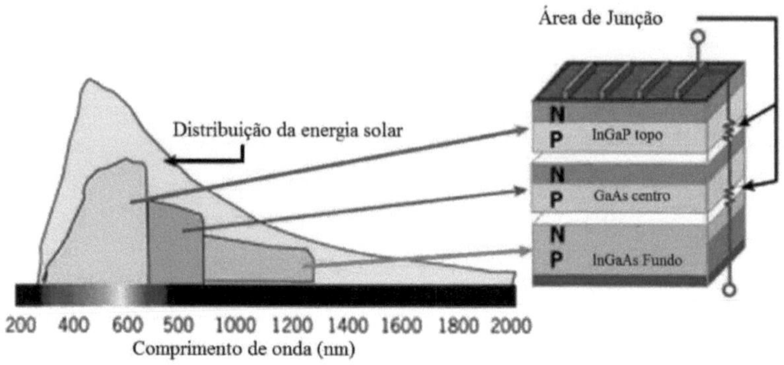

Figure 7 - Multi-junction solar cell with three pn junctions

Source: [16]

The most efficient laboratory-produced multi-junction cells are triple junction cells with an efficiency of 37.9%, while organic and dye-sensitised cells have a current record of 10.7% and 11.9% respectively [16, 17].

2.3 CHARACTERISATION OF PHOTOVOLTAIC CELLS

Solar cells can be characterised optically and electrically. Optical characterisation is done using a spectrophotometer, which shows the region of greatest light absorption for a given cell material, while electrical characterisation is most often done using solar simulation equipment [7].

The electrical characterisation of solar cells is essential in order to make them more efficient, with minimal losses [7]. In this section, only the electrical characterisation that is the focus of this work will be presented.

2.3.1 I-V curve

In order to assess the performance of a solar cell, it is necessary to obtain its characteristic curve (I-V curve), from which it is possible to derive important parameters for assessing the best operating mode for the system. By varying the resistive loads on the terminals of these cells, different values can be obtained for the voltage and current, which when plotted on the V x I graph shows the characteristic curve. Each operating point has its own specific voltage and current, and the product of these values gives the power [7, 17]. Figure 8 illustrates the I-V curve and shows how the parameters short-circuit current (Isc), open-circuit voltage (Voc), maximum power current (Imp), maximum power voltage (Vmp), maximum power (Pmax) and fill factor (FF) are obtained graphically [16].

Figure 8 shows the following quantities (which are important for characterising and operating the cells):

- Short-circuit current (Isc): this is the current when the voltage at the cell terminals is zero or closed (zero resistive load - short-circuited), being the point of intersection between the curve and the current axis. In this case (see Figure 10 which shows the equivalent electrical circuit in the next section[2]), the voltage across the diode is equal to the voltage across the series resistor Rs. However, Rs has a low resistance value, so the voltage across Rs and the diode is small and insufficient for the diode to start conducting.

As the resistance Rsh is very high, all the current flows through the short-circuited path. The short-circuit current can be described in terms of the short-circuit current density Jsc times the area A of solar incidence (equation (1)). Jsc values vary according to the material and technology used [7, 17].

[2] The equivalent electrical circuit is used to obtain the characteristic curve as discussed in the next section. However, it is necessary to know the I-V curve first before understanding the circuit. To understand the parameters and equations in this section (2.3.1), we recommend reading section 2.3.2.

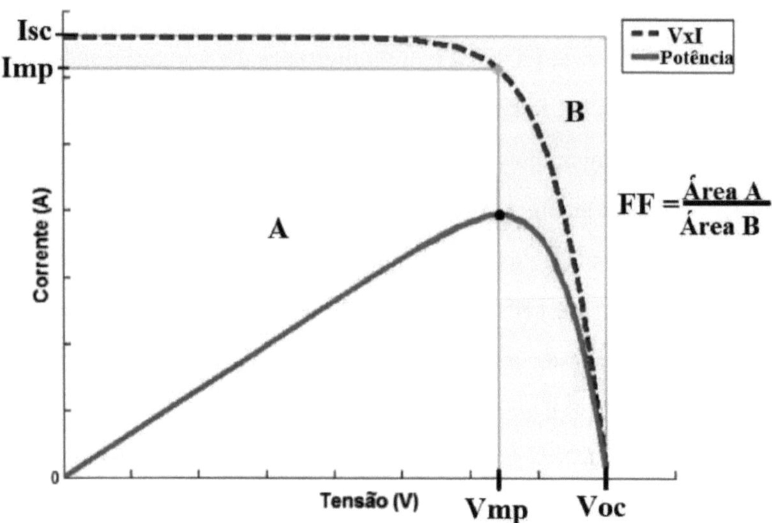

Figure 8 - Characteristic curve of solar cells

Source: [7]. Adapted.

$$Isc = Jsc.A \quad (1)$$

- Open circuit voltage (Voc): electrical potential difference when the terminals of the cell are open, i.e. an infinitely high resistive load is connected to its terminals, and all the current generated flows through the diode, so the output current is zero at Voc. In this case, Voc also depends on the material and manufacturing technique used. It can be calculated, after some manipulations of equations (6) and (7) in the next section, as equation (2), which is valid for irradiation G = 1000 W/m^2 . Where Is is the reverse saturation current, n is the ideality factor, k = 1.38 x 10^{23} J/K is the Boltzmann constant, q = 1.602 x 10^{-19} C is the electron charge, Tc is the cell temperature and Isc is the short-circuit current [7, 17].

$$Voc = \frac{nkT_C}{q} \ln\left(1 + \frac{I_{SC}}{I_S}\right) \quad (2)$$

-Maximum power point (Pmax): Place on the curve where the product of voltage and current is maximum; it represents the maximum operating power point, where the voltage and current values are called maximum power voltage Vmp and maximum power current Imp [17]. It is obtained from the area A in

Figure 8 or from equation (3) [16].

$$P_{max} = V_{mp} I_{mp} \tag{3}$$

-Fill factor (FF): Figure 8 also shows the fill factor, which is another parameter for evaluating the cell's performance; FF can be calculated using equation (4) (equation (4) is obtained by dividing area A by area B of the graph in Figure 8). FF values greater than 0.7 represent high quality cells (first generation and multi-junction cells have fill values between 0.8 and 0.9); second line cells have a factor between 0.4 and 0.7, representing lower performance. This magnitude also depends on the material and manufacturing technique [17].

$$FF = \frac{Vmp\ Imp}{Voc\ Isc} \tag{4}$$

AND Conversion efficiency or yield: this can be calculated as the ratio between the maximum power Pmax and the incident power Pin (Pin is obtained by the solar radiation G times the cell's area A), according to equation (5) [16].

$$\eta = \frac{Pmax}{Pin} = \frac{Voc\ x\ Isc\ x\ FF}{G\ x\ A} \tag{5}$$

2.3. 2Equivalent Electrical Circuit

To study the physical parameters of how photovoltaic cells work, it is common to use equivalent circuit models to simulate and analyse the I-V curve. The solar cell, under ideal conditions, can be represented by a current source in parallel with a diode as shown in Figure 9. The current generated by solar irradiation is shown as IL, the current in the diode is described by ID, while at the terminals of the cell there is the net current Ic (cell current) and the voltage Vc (cell voltage). According to Kirchhoff's law of nodes [19], the current Ic is obtained from equation (6) [6,5].

$$Ic = I_L - I_D \tag{6}$$

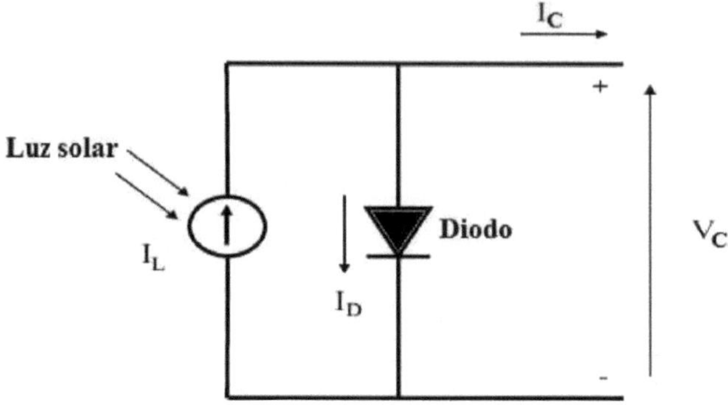

Figure 9 - Model of a diode for an ideal cell

Source: [5]. Adapted.

ID can easily be obtained from the Shockley equation for real diodes (equation (7)). Where Is is the reverse saturation current of the diode, VD is the voltage on the diode, n is the ideality factor (n = 1 represents ideal behaviour of the diode), k = 1.38 x 10^{23} J/K is the Boltzmann constant, q = 1.602 x 10^{-19} C is the electron charge, Tc is the temperature of the diode (as the diode represents the solar cell then the temperature of the cell is the same as the diode) [15].

$$I_D = I_S \left[e^{\left(\frac{qV_D}{nk\,T_c}\right)} - 1\right] \tag{7}$$

IL is directly proportional to the irradiance received by the photovoltaic cell, and can be obtained from equation (8) where Jsc is the short-circuit current density (mA/cm^2), A is the area of the photovoltaic cell (cm^2) and G is the irradiance (W/m^2) [5].

$$I_L = \frac{J_{SC} A \text{ x } G}{1000} \tag{8}$$

Substituting equations (7) and (8) into (6) and considering G = 1000 W/ m^2 gives equation (9) [5].

$$I_c = J_{sc}A - I_S\left[e^{\left(\frac{qV_D}{nk\,T_c}\right)} - 1\right] \tag{9}$$

Substituting (1) into equation (9), and as the circuit is in parallel, you can also substitute the voltage on the diode VD for the cell voltage Vc (see Figure 9):

$$I_c = I_{sc} - I_S\left[e^{\left(\frac{qV_C}{nkT_c}\right)} - 1\right] \quad (10)$$

However, the model in Figure 9 and equation (10) represents ideal cells, and in practice, the model that best represents photovoltaic cells has series resistance (Rs) and parallel resistance (Rsh) (Figure 10), which characterise losses due to cell failures caused by manufacturing problems and operating conditions (see section 2.3.3), and this circuit is called the One Diode Model [6]. Rs appears due to voltage loss at the metal contacts and ohmic losses on the front surface of the cells, as well as impurities, and under ideal conditions should tend to zero, not allowing the cell voltage to drop in Rs. Rsh is related to defects in the crystalline structure, under ideal conditions it should tend to infinity, not allowing current to pass through [7].

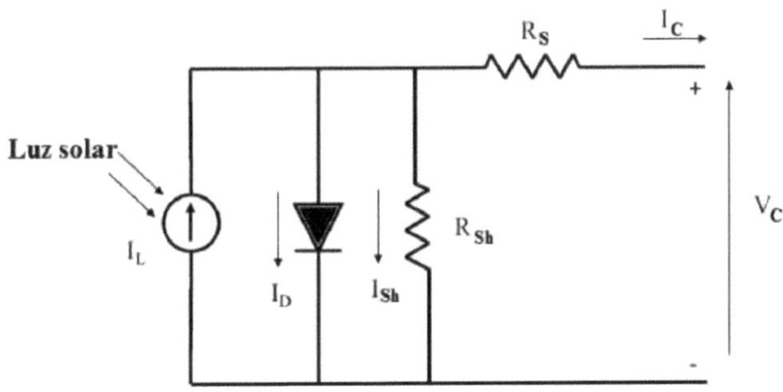

Figura 10 - **Model of a Diode**

Source: [5]. Adapted.

High quality cells have low resistance Rs and high resistance Rsh. The effects of the resistances on the I-V Curve can be seen in Figure 11 [17].

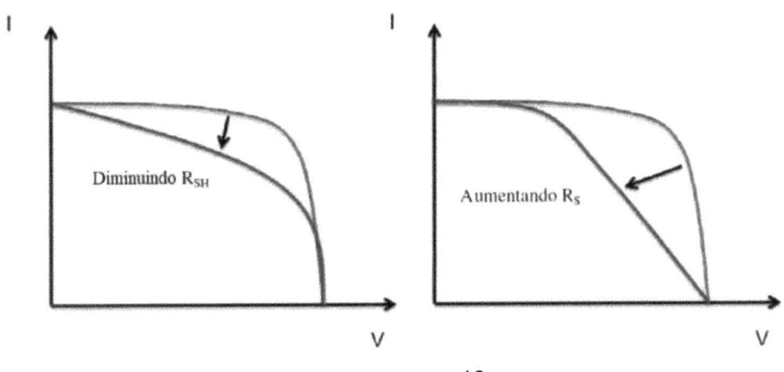

Figura 11 - Rsh and Rs resistance effect

Source: [16]

It can be seen that as the series resistance increases and the shunt (or parallel) resistance decreases, the maximum power and fill factor decrease, causing the cell to lose performance [5, 16].

Applying Kirchhoff's law of nodes [19] to the circuit in Figure 10, the current Ic is obtained from equation (11). Where Ish is the current that arises in the parallel (shunt) resistor, and can be obtained from equation (12). The potential difference across the diode and shunt resistor is given by equation (13) [5, 16].

$$I_C = I_L - I_D - I_{sh} \tag{11}$$

$$I_{Sh} = \frac{V_c + I_c R_s}{R_{sh}} \tag{12}$$

$$V_D = V_c + I_c R_s \tag{13}$$

The model in Figure 10 is currently the most widely used and can be modelled mathematically by equation (14), where Vt is given by equation (15) [5].

$$I_C = I_L - \mathrm{I_S}\left[\mathrm{e}^{\left(\frac{V_c + I_c R_s}{V_t}\right)} - 1\right] - \frac{V_c + I_c R_s}{R_{sh}} \tag{14}$$

$$V_t = \frac{nkT_c}{q} \tag{15}$$

Although the One Diode Model can achieve acceptable levels of accuracy, the most accurate representation of the photovoltaic cell can be made by two Shockley diodes in parallel with a current source and associated resistors, Rs and Rsh. See the model in Figure 12 called the Two-Diode Model [5, 6].

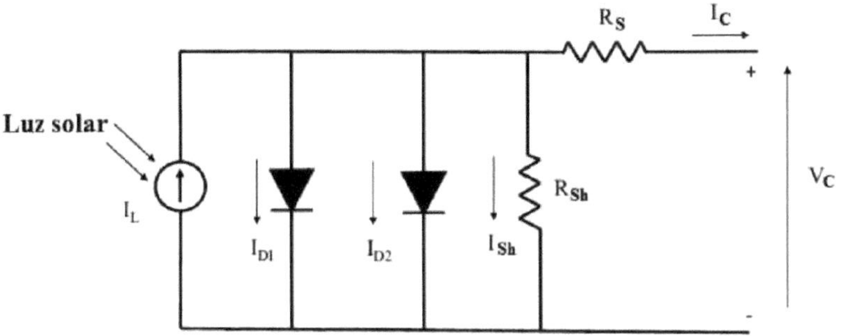

Figura 12 - Two-Diode Model

Source: [5]. Adapted.

This model (Figure 12) better characterises the solar cell at the low level of operation with low solar irradiance, and allows for better adjustment due to the effect of recombination [5]. Electrons in the conduction band can recombine with oxidised impurities and return to the valence band of the semiconductor, causing losses in electricity generation. This effect is called recombination. In Si cells, electrons are lost when they recombine with defects and impurities in the crystal lattice [21].

The mathematical equation for the equivalent circuit of the two-diode model is given by equation (16), which is obtained in a similar way to the one-diode model. For simulation and simplification purposes, the ideality factor n (equation (15)) is taken, n = 1 for D1 (Diode 1) and n = 2 for D2 (Diode 2), where 1 < n < 2 [5].

$$I_C = I_L - I_{S1}\left[e^{\left(\frac{V_C + I_C R_S}{V_{t1}}\right)} - 1\right] - I_{S2}\left[e^{\left(\frac{V_C + I_C R_S}{V_{t2}}\right)} - 1\right] - \frac{V_c + I_c R_s}{R_{sh}} \qquad (16)$$

2.3.3 Operating conditions

Climatic conditions influence the performance of photovoltaic cells, impacting on their ability to generate electricity. The main factors are light intensity and temperature, which depend on solar spectral distribution [17, 18].

2.3.3.1 Spectral distribution and air mass

The spectral distribution is attenuated when it interacts with the atmosphere through the absorption and scattering of solar radiation. Absorption occurs through the loss of solar energy to atoms and molecules in the environment, which can be dissipated in the form of surrounding heat. Scattering causes reflection of electromagnetic waves, reducing the incidence of solar radiation. The way these phenomena occur depends on the physical characteristics of the atmosphere. The intensity of these two processes will depend on the composition and distance travelled by the radiation in the atmosphere. The air mass (AM) is precisely this distance, which depends on the zenith angle, which varies with latitude, time of day and day of the year. See the relationship between AM and the zenith angle in Figure 13 [17, 22].

AM 0 represents free space, i.e. above the atmosphere. AM 1.5 is used in tropical regions, AM 1.0 is used at the equator, and high latitude regions use AM 2.0 or AM 3.0. The value of AM can be obtained from equation (17), which considers the atmosphere to be homogeneous and non-refractive, as well as disregarding the curvature of the Earth, so its **use is restricted to small zenith angles (10% error for 0z = 80°). Where 0z** is the zenith angle [17, 22].

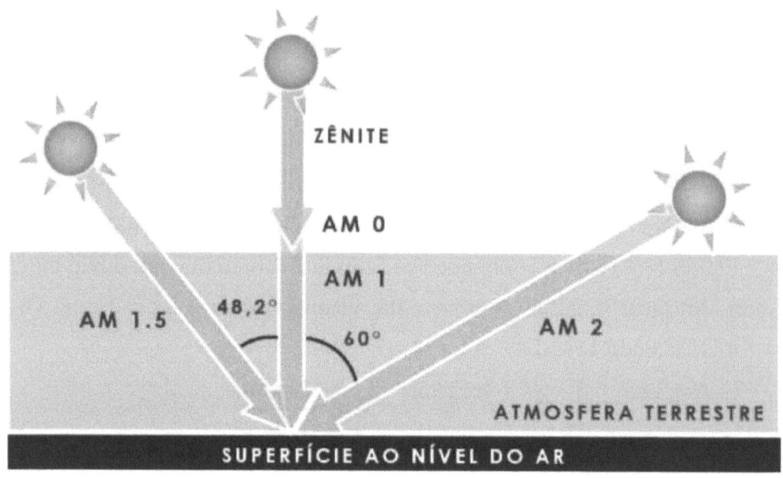

Figure 13 - Air mass in relation to the zenith angle

Source: [22]

$$\mathrm{AM} = \frac{1}{\cos(\theta_z)} \quad (17)$$

As a result, the spectral distribution has a direct impact on the IL current, because the type of semiconductor material used to manufacture the cells can respond to certain wavelength ranges of solar irradiation differently, i.e. there is a spectral response for each cell made of different materials. For example, silicon generates useful current in the 40 nm to 900 nm range [17].

Figure 4 shows the spectral distribution for some components of the atmosphere where the intensity of the radiation depends on the wavelength. Gases in the atmosphere reduce the solar spectrum at points on the AM1.5 curve (note the falling peaks in Figure 4), making it lower than the spectrum outside the atmosphere (AM 0). On cloudy days, the curve in red, the spectral distribution is low on the Earth's surface, which further reduces the IL current [7].

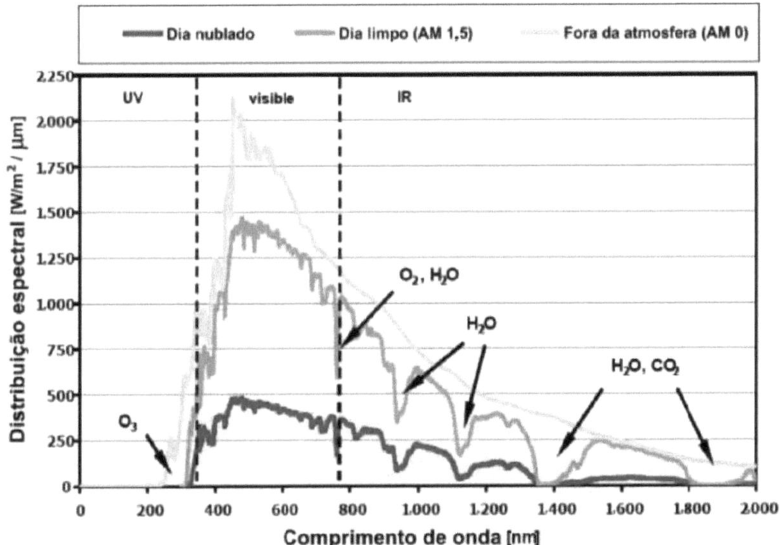

Figure 14 - Spectral distribution of solar radiation

Source: [7]

2.3.3. 2Irradiance

The power incident per unit area on the surface G is called solar irradiance, and is understood as the area (integral) of the spectral distribution. It can be related to the air mass with the approximate relationship in equation (18) [8].

$$G = G_0 e^{-0,357xAM^{0,678}} \quad (18)$$

Where G0 is the solar constant, i.e. the irradiance measured outside the atmosphere, and has a value equal to 1367 W/m^2 . So the air mass will depend on the region of the planet [17].

The irradiance has an impact on the maximum power supplied. Figure 15 shows the characteristic curve for different values of irradiance at the same temperature, where there is less power at low values of solar radiation, with Isc falling dramatically in relation to the decrease in Voc [23].

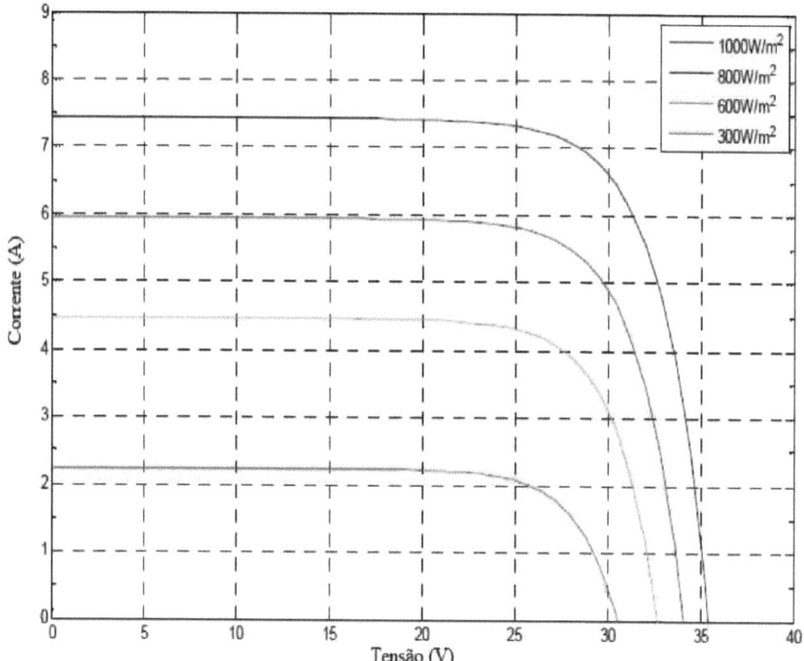

Figure 15 - I-V curves for different irradiances

Source: [23]. Adapted.

2.3.3. 3Temperature

The ambient temperature and the level of incident solar radiation have an impact on the temperature of the cells that make up the solar panel. Figure 16 shows the characteristic curves for different operating temperatures. The increase in temperature characterises a decrease in the Voc voltage and a slight increase in the Isc current, causing a change in the power delivered by the module, which will be greater at lower temperatures. When compared with the irradiance, the opposite can be seen, as higher irradiance values characterise greater maximum power delivered by the cell

[12, 23].

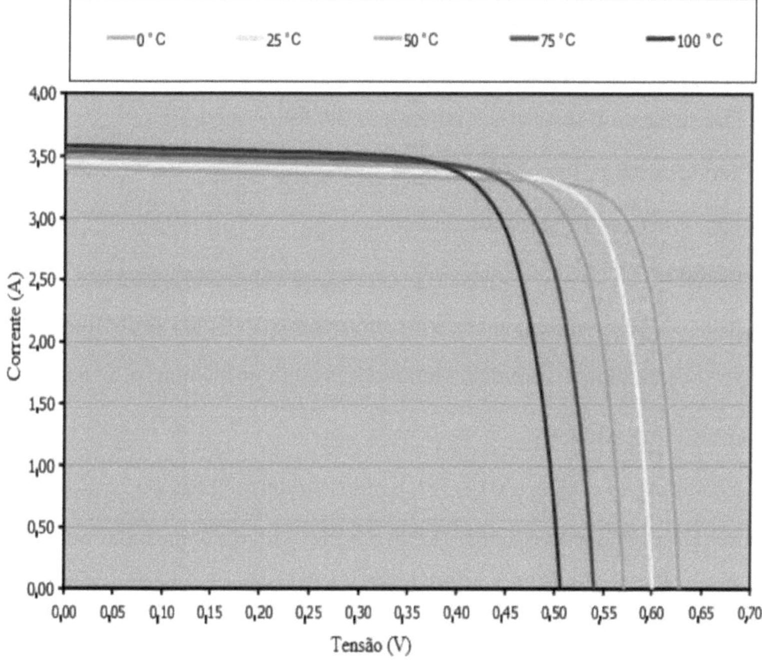

Figure 16 - I-V curves for different temperatures

Source: [17]. Adapted.

2.3.3. 4Standard Test Condition - STC

Solar module manufacturers provide electrical characteristics under standard test conditions (STC). For this operational analysis, well-defined environmental parameters are used, which are irradiance = 1000 W/m^2 , air mass AM = 1.5 and temperature = 25 °C [17].

CHAPTER 3

MATERIALS AND METHODS

3.1 SIMULATION METHODOLOGY

The simulation in Ltspice[3] comprises the following steps:

1. Choice and creation of the Ltspice model for the simulation;
2. Vary the parameters of the model created;
3. Survey of the I-V curve;
4. Comparison of the simulated curve with photovoltaic cell curves obtained analytically; 5. Extraction of simulated parameters and verification of data reliability;

3.2 BUILDING THE MODEL

According to Sarkar (2016) [5], a photovoltaic cell can be represented by a current source (Iph), one or two diodes (D) in parallel with the source and two resistors, one in series (Rs) and one in parallel (Rp). Therefore, in order to apply the methodology, firstly, using the LTspice software, equivalent circuit models were built for commercial silicon solar cells, based on this description.

Figure 17 shows the simplest model of a diode made in LTspice. It also shows a voltage source V1 that varies between a range of 0 and 20 V. This voltage variation (which acts as a resistive load) is necessary to obtain the I-V curve of the simulated solar cell.

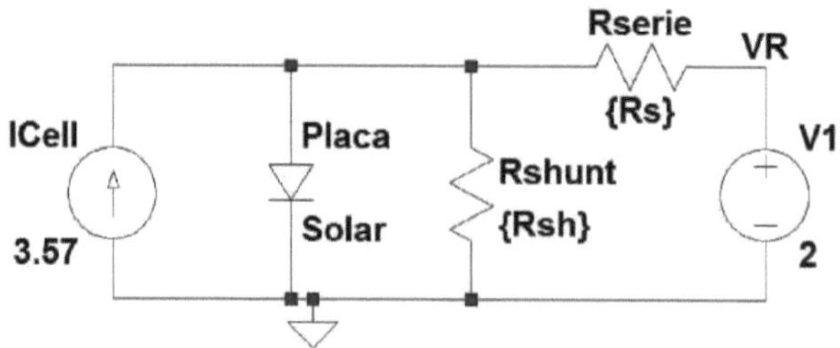

Figure 17 - Simplest model of a diode

[3] LTspice ® is a computer program for simulating analogue circuits, produced by Linear Technology [24].

Source: Own authorship.

The one-diode model achieves an acceptable level of precision, but the most accurate electrical representation of the solar cell can be made by two Shockley diodes in the previous circuit [5].

A more realistic model was developed from a tutorial [25] to obtain the characteristic curves (Figure 18). This circuit has two current sources, Icell and Fcell, where Fcell is a current-dependent current source, two diodes, D1 and D2, the series and parallel resistors, Rserie and Rshunt respectively, two voltage sources Epv and V1, where Epv is a voltage-dependent voltage source, the resistor R1 only serves to make the circuit more real, since both voltage sources cannot be connected directly in parallel. Here V1 also varies over a range from 0 to 20 V. This model could be built without Fcell and Epv, however, it serves to adjust the curve depending on the number of cells in series and parallel, i.e. changing the current and voltage of the solar module.

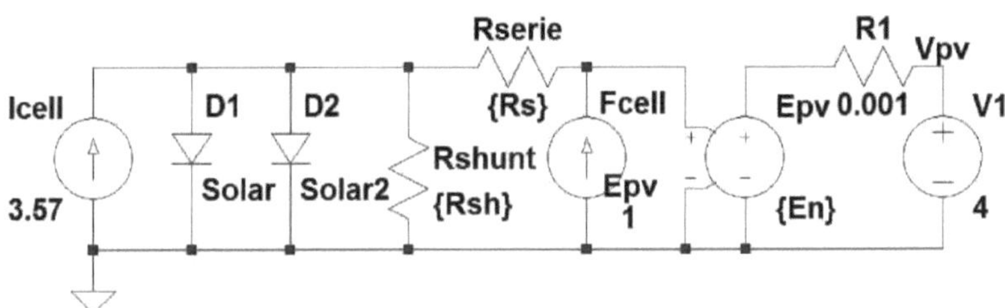

Figure 18 - More elaborate model of two diodes in parallel

Source: Own authorship.

3.3 SIMULATION

The characteristic curves of the photovoltaic cell from the article by GNOATTO, et al. (2005) [2] were used for simulation. In this article, the authors surveyed two I-V curves under real working conditions (experimentally), where they obtained Figure 19 and Figure 20. Figure 19 shows a maximum power voltage (Vmp) of 13.34 V, a maximum power current (Imp) of 3.23 A, an open circuit voltage (Voc) of 18.74 V, and a short circuit current (Isc) of 3.57 A, with a maximum power (Pmax) of 43.14 W and an efficiency of 9.57%. To obtain Figure 20, the same photovoltaic panel was used, but a continuous water film was applied to reduce the panel's operating temperature (around 35 °C), resulting in an increase in maximum power and efficiency of the characteristic curve. In this case, they obtained a Vmp of 14.54 V, Imp of 3.22 A, Voc of 20.16 V, Isc of 3.50 A, with a Pmax of 46.75 W and an efficiency of 10.36%.

The Ltspice simulation was then carried out to obtain the I-V curve equivalent to Figure 19 and Figure 20. The parameters of the circuit created in Figure 18 were altered to obtain the simulated curves and find the parameters that describe the panel.

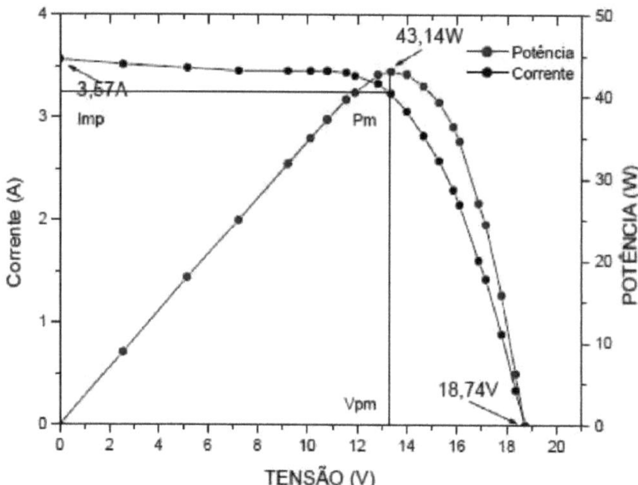

Figure 19 - I-V characteristic curve under field conditions at 54 °C

Source: [2]

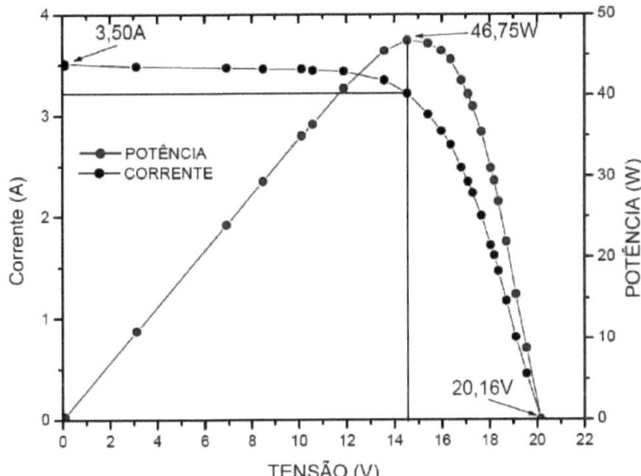

Figure 20 - I-V characteristic curve under field conditions at 35 °C

Source: [2]

CHAPTER 4

RESULTS AND DISCUSSION

4.1 EFFECT OF VARYING PARAMETERS

This section will discuss and evaluate the impact of the parameters in the two-diode model on the operation of photovoltaic cells. The parameters discussed are the resistances Rs (series) and Rsh (parallel), the saturation currents Is of D1 and D2, the effect of the current IL of IC and the voltage Epv.

4.1.2 Effect of resistance Rs and Rsh

Before simulating the characteristic curve in Figures 19 and 20, the parameters of the model created were varied in order to check the effects they have on the solar cell's I-V curve. Firstly, the series resistance Rs was varied, keeping the resistance Rsh constant; Figure 21 shows the result of the effect of Rs through the graph of the I-V curve. The software used to create the graphs was QtGrace .[4]

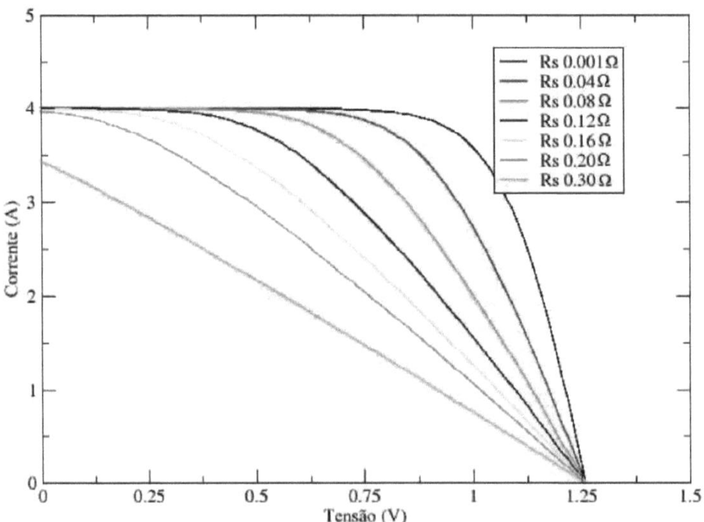

Figure 21 - Graph of the effect of varying the series resistance Rs

[4] QtGrace / Grace is a program for displaying or plotting data, analysing data and preparing it for printing [26].

Source: Own authorship.

As can be seen in Figure 21, with the increase in series resistance Rs, the open-circuit voltage Voc (1.25 V) remains the same, but the maximum power and the fill factor FF are significantly reduced. In the simulation, the value of the series resistor Rs is kept quite low to avoid a voltage drop in the series resistor, which causes degradation in the performance of the curve. Panels with good performance have a characteristic curve similar to red or green, with the black curve being an ideal panel with maximum performance.

Figure 22 shows the effect of varying the parallel resistance Rsh (Rp in the graph). When making this sweep, it can be seen that the value of the short-circuit current Isc (4 A) remains the same, but the open-circuit voltage Voc and the fill factor FF reduce dramatically as the resistance of Rp (Rsh) decreases. Thus, the shunt (parallel) resistance is kept high (100 **Q in this example).** Another important point is that the curve becomes linear, obeying OHM's law, at low values of Rp, because the current passes completely through the resistors as if there were no diodes, since Rp is very low.

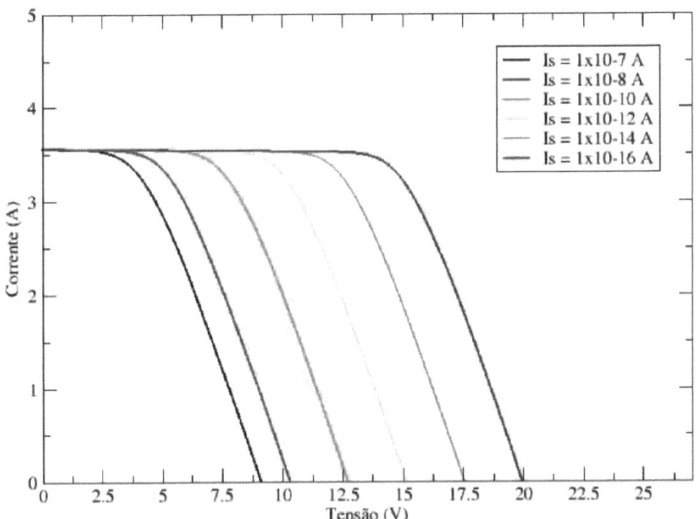

Figure 22 - Effect of varying the resistance in parallel Rsh

Source: Own authorship.

4.1.2 Effect of saturation current on D1 and D2

The impact of the reverse saturation currents of diodes D1 and D2 is shown in Figure 23 and Figure 24. It is clear that, in both cases, the short-circuit current Isc remains the same, but the open-circuit voltage Voc and the fill factor FF decrease. As the saturation current decreases, the potential difference across the open circuit Voc increases, and so FF increases. This is because the Shockley equation for a diode, equation (7), depends on the exponential function [27], as this function has the characteristic of changing its slope as e^x is multiplied by a constant, where Is plays the role of the multiplied constant.

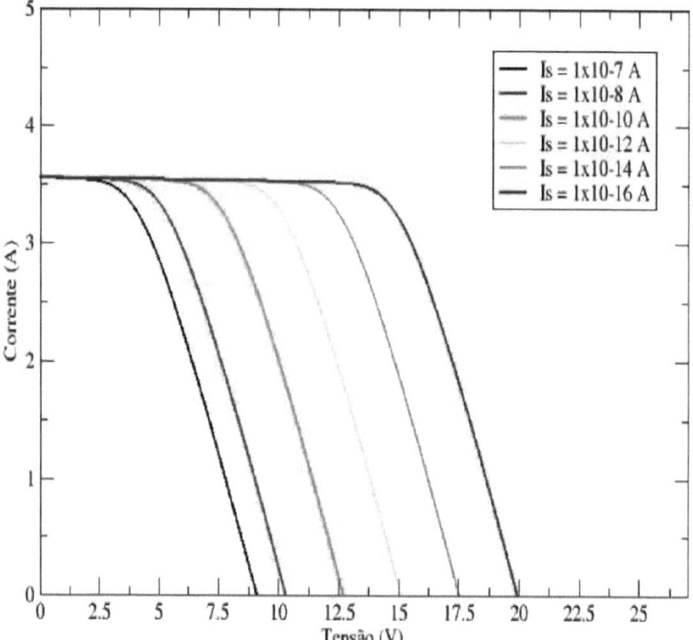

Figure 23 - Effect of varying the saturation current (Is) of D1

Source: Own authorship.

In Figure 23, lower values of Is will cause the diode to start conduction at higher voltage values, which is consistent with equation (16); the point at which the diode starts conduction is precisely when the cell current drops. As a result, the potential difference in the Voc open circuit is greater. In Figure 24 the process is the same, but for very small values of Is, the current in diode D2 will be very small until it has no influence on the circuit; note that for values less than 1 x 10^{-10} A, the characteristic curve of the solar cell does not change.

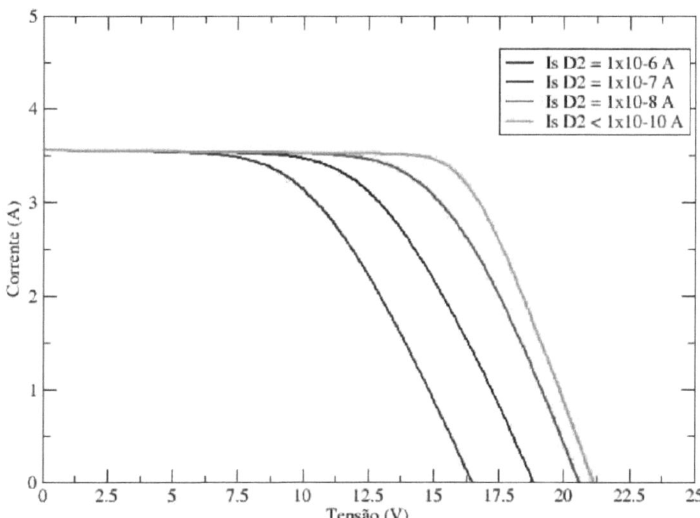

Figure 24 - Effect of varying the saturation current (Is) of D2

Source: Own authorship.

4.1.4 Effect of other parameters in simulation

Other parameters that could be varied are the current IL (value of the current source) and the voltage Epv (value of the voltage-dependent voltage source) in the model in Figure 18. These parameters directly influence the short-circuit current Isc and the open-circuit voltage Voc, respectively (Figure 26 and Figure 27). In addition, Ltspice is a complete ferment, other parameters that have not been mentioned can be modified to improve the model, for example, the ideality factor n, the diode temperature, the activation energy EG, the junction potential pn, among other diode coefficients. However, these parameters are not relevant to this simulation.

In Figure 26, the IL value is directly linked to the irradiance received by the cell, where the greater the irradiance G, the greater the IC current generated in the solar cell. Note that the short-circuit current Isc is the same as IL, because a current source was used to simulate the irradiance. In this case, there are no significant changes in Voc, which depends on the diode's saturation current, which remains the same.

Varying the value of Epv in Figure 27 shows an increase in Voc and consequently an increase in maximum power. Increasing Epv is equivalent to increasing the number of cells in series, i.e. replicating the model in series, so the final voltage increases but the Isc current remains the same.

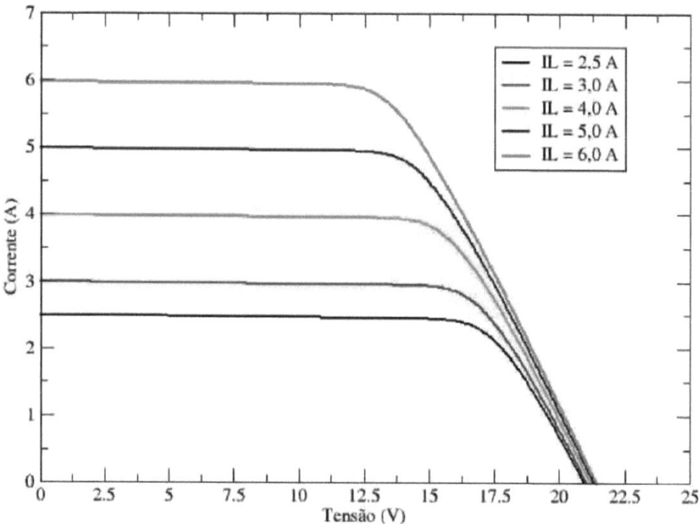

Figure 25 - Effect of varying the current (Il) on the model

Source: Own authorship.

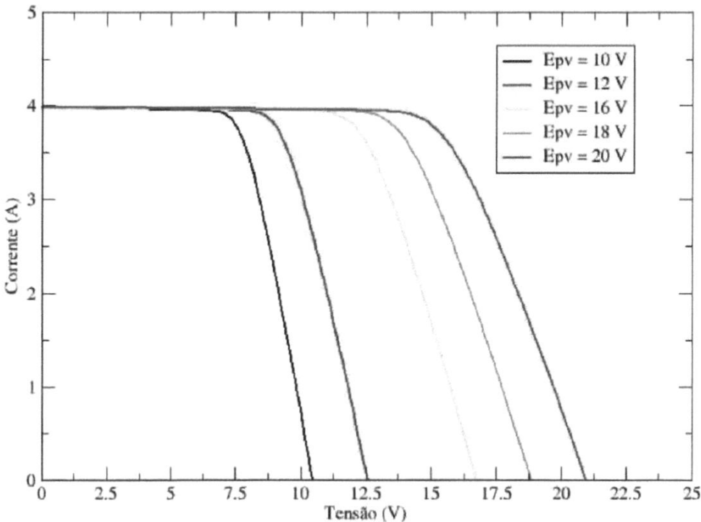

Figure 26 - Effect of varying the potential difference (Epv) on the model
Source: Own authorship.

4.2 ADJUSTING THE I-V CHARACTERISTIC CURVE

After analysing the variation in parameters, the characteristic curve is obtained by simulating it in Ltspice. Figure 28 and Figure 29 show the simulated curves after some parameter adjustments in the software. We tried to obtain the same data as GNOATTO, et al. (2005) [2] as described in the methodology.

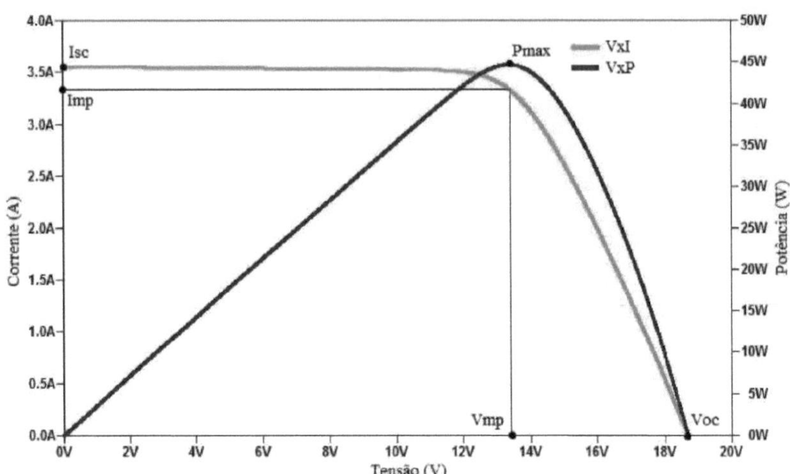

Figure 27 - I-V characteristic curve of the first simulation

Source: Own authorship.

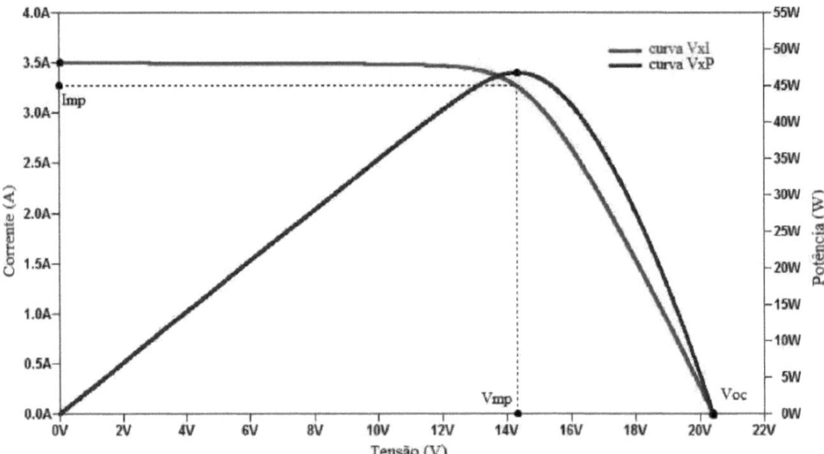

Figure 28 - I-V characteristic curve of the second simulation
Source: Own authorship.

In Figure 28, the green curve represents the voltage per current and the blue curve represents the voltage per power, where the parameters removed after simulation and relative percentage errors can be seen in Table 1. Experimentally, it is not possible to obtain the saturation currents of the diodes, nor the series and parallel resistance, so a dash was placed on these parameters provided by the authors [2].

GNOATTO, et al. (2005) [2] used two panels in parallel with polycrystalline silicon cells connected in series, which could supply up to 112.2 W and an efficiency of approximately 12% for both, under STC conditions. The article's experimental values under real working conditions for the first curve can also be seen in Table 1.

Table 1 - Data from the first simulation

Greatness	Article value [2]	Simulated value	Relative error
You	18,74 V	18,70 V	0,21 %
Isc	3,57 A	3,56 A	0,28 %
Vmp	13,34 V	13,36 V	0,14 %
Imp	3,23 A	3,34 A	3,40 %
Pmax	43,14 W	44,60 W	3,38 %
FF	0,64	0,67	4,68 %
Is D1	-	$1x10^{-15}$ A	-
Is D2	-	$1x10^{-16}$ A	-
Lol	-	0,056 Ω	-
Rsh	-	18 Ω	-
Income	9,56 %	9,89 %	3,45 %

Source: Own authorship.

According to Table 1, the values were close because the relative error was low, so the first curve was completely characterised.

In the second simulation, the parameters taken from Figure 29 and the relative percentage errors can be seen in Table 2, where the red curve represents the voltage per current and the blue curve represents the voltage per power.

Here the values were closer than in the first simulation because the relative errors for each parameter were smaller. It is important to note that the same panel was used to obtain the characteristic curves by GNOATTO, et al. (2005) [2], only the operating temperature of the solar panel was changed. This influenced the gain in performance and consequently the model resistances improved in value.

Table 2 - Parameters obtained from the second simulation

Greatness	Article value [2]	Simulated value	Relative error
You	20,16 V	20,40 V	1,19 %

Isc	3,50 A	3,50 A	0 %
Vmp	14,54 V	14,30 V	1,65 %
Imp	3,22 A	3,27A	1,55 %
Pmax	46,75 W	46,74 W	0,02 %
FF	0,66	0,65	1,51 %
Is D1	-	$1x10^{-17}$ A	-
Is D2	-	$1x10^{-9}$ A	-
Lol	-	0,069 Ω	-
Rsh	-	38 Ω	-
Income	10,36 %	10,36 %	0 %

Source: Own authorship.

Anyway, analysing the two simulations, Rs and Rsh, which are the main parameters for characterisation, had a reasonably better value than the first case, in fact, Rs changed a little and Rsh increased. As a result, Pmax increased and FF remained at almost the same value (this was because it was the same cell at a different temperature). However, in the second case FF should have been higher when compared with the data in the article in Table 1 and Table 2, but the values are close and have a small relative error. The gain in maximum power can be attributed to the shunt resistance, which increased by approximately 20 Ω (improving performance). As previously mentioned, high-performance cells have high Rsh values and very small Rs values. Rs increased slightly, but not enough to compromise performance.

It is worth noting that in this model, the temperature of the diodes was not varied, which could improve the simulation in terms of relative errors. Thus, with these parameters, all that is needed is to improve performance and make it feasible to apply the panel in a given location, and in both cases the STC standard was not used. As UFVJM does not yet have any silicon photovoltaic panels to carry out experiments in the field, this work was limited to analysing the characteristic curves of the solar cells in the article through simulation.

CHAPTER 5

CONCLUSION

The aim of the work was achieved, since the equivalent electrical circuit simulation provided the parameters needed to analyse the characteristic curves of the simulated cell. This analysis basically consists of determining the losses, i.e. the Rsh and Rs values, which determine how much the cell loses of its ideality, and can also be compared with the STC standard, which represents the best way of operating, which verified the percentage of loss of performance in conditions that are not so varied. However, other parameters have also been determined and can be used to evaluate solar cell constructions. The aim of this work is to simulate and provide the parameters needed to evaluate the cells. In Ltspice, these parameters basically influenced the conversion efficiency and the fill factor.

As future prospects, this study can be used to improve the construction of the silicon cells tested in the laboratory, and can also be adapted to determine the parameters of organic and dye-sensitised cells, which are still under development. Thus, the simulation analysis is promising and the model can be improved, for example, by replacing the current source with a more real component that takes into account the solar absorption spectrum, the cell temperature, among other parameters.

CHAPTER 6

BIBLIOGRAPHICAL REFERENCES

[1] OLIVEIRA, A. F.; FERREIRA, G. M. **Study and simulation of the optical performance of solar cells.** II Simpósio de Iniciação Científica da Universidade Federal do ABC, SIC-UFABC, Santo André - SP, 2009.

[2] GNOATTO, E. et al. **Determination of the characteristic curve of a photovoltaic panel under real working conditions.** Acta Sci. Technol. Maringá, v. 27, n. 2, p. 191-196, July/Dec., 2005.

[3] EARTH. **Solar energy achieves historic growth in Brazil.** Financing of R$3.2 billion should further accelerate the sector. 23 May 2018. Available at: <http://www.absolar.org.br/noticia/noticias-externas/energia-solar-fotovoltaica.html>. Accessed on: 11 August 2018.

[4] BRAZILIAN PHOTOVOLTAIC SOLAR ENERGY ASSOCIATION (ABSOLAR). **Photovoltaic solar energy.** Available at: <http://www.absolar.org.br/noticia/noticias-externas/energia-solar-fotovoltaica.html>. Accessed on: 09 Apr. 2018.

[5] SARKAR, Md. N. I. **Effect of various model parameters on solar photovoltaic cell simulation**: a SPICE analysis. Institute of Energy, University of Dhaka, Bangladesh. SpringerOpen Journal, 2016.

[6] FARIAS, J. L. C.; MALAGÓN, L. A. G. **Simulation of photovoltaic solar cells using NI MULTISIM software. COBENGE**, XLI Brazilian Congress of Engineering Education. Gramado - RS, 2013.

[7] FREITAS, D. H. **Installation and operation of a solar simulator and tests with solar cells.** 2015. 83 f. Course Conclusion Work (Undergraduate Degree in Electrical Engineering) - Department of Electrical Engineering, University of Brasília, Brasília - DF, 2015.

[8] GARUZZI, R. P.; ROMERO, O. J. Economic feasibility of implementing photovoltaic cells in homes in Espirito Santo, Brazil. **Revista ESPACIOS,** v. 38, n. 1, p. 23, 2017.

[9] DAY TO DAY EDUCATION. **Doping - Types of Doping - Chemical Doping of Materials - How Doping Occurs in Semiconductors.** Available at: <http://www.quimica.seed.pr.gov.br/modules/conteudo/conteudo.php?conteudo=175 >.
Accessed on: 12 August 2018.

[10] MATAVELLI, A. C. **Solar energy: generating electricity using photovoltaic cells.** 2013. 34 f. Final Course Work (Undergraduate Degree in Chemical Engineering) - Lorena School of Engineering, University of São Paulo, Lorena - SP, 2013.

[11] SHUKLA, A.; KHARE, M.; SHUKLA, K. N. **Modelling and Simulation of Solar PV Module on MATLAB/Simulink.** International Journal of Innovative Research in Science, Engineering and Technology - IJIRSET. Vol. 4. India, 2015.

[12] FADIGAS, E. A. F. A. **Photovoltaic Solar Energy: Fundamentals, Conversion and Technical and Economic Viability.** PEA - 2420- ENERGY PRODUCTION. GEPEA - Energy Group. Polytechnic School, University of São Paulo, year not specified.

[13] PEREIRA, L. S. INFO ESCOLA NAVEGAR E APRENDENDO. **Electronic doping**. Available at: < https://www.infoescola.com/quimica/dopagem-eletronica/> Accessed on: 12 Aug. 2018.

[14] MANZALE, G. Electro Physics - Diode. Available at: < http://eletrofisica1.blogspot.com/2016/10/diodo-semicondutor.html> Accessed on: 12 Aug. 2018.

[15] LUQUE, A.; HEGEDUS, S. **Handbook of photovoltaic science and engineering**. 2. ed. Publisher: Wiley, Copyright ©, 2003.

[16] SILVA, T. F. P. **Application of carbon nanostructures in organic and inorganic solar cells.** 2015. 113 f. Thesis (Doctorate in Electrical Engineering) - School of Electrical

and Computer Engineering, State University of Campinas, Campinas - SP, 2015.

[17] CARVALHO, André L. Costa. **Methodology for analysing, characterising and simulating photovoltaic cells.** 2014. 96 f. Dissertation (Master's in Electrical Engineering) - Department of Electrical Engineering, Federal University of Minas Gerais, Belo Horizonte - MG, 2014.

[18] BRAGA, R. P. **Photovoltaic solar energy: fundamentals and applications.** 2008. 80 f. Course Conclusion Paper (Undergraduate Degree in Electrical Engineering) - Department of Electrical Engineering, Federal University of Rio de Janeiro, Rio de Janeiro - RJ, 2008.

[19] HALLIDAY, David; RESNICK, Robert; WALKER, Jearl. **Fundamentals of Physics - Electromagnetism**. 10th edition, vol. 3, LTC. 2016.

[20] BOYLESTAD, R. L.; NASHELSKY, L. **Electronic Devices and Circuit Theory.** 11. ed. Publisher: Pearson, 2013. 784 p.

[21] SANTOS, A. J. **Separation and Recombination of Charges in Photoelectrochemical Solar Cells.** 2010. 158 f. Thesis (Doctorate in Condensed Matter Physics) - Institute of Physics, Federal University of Alagoas, Maceió - AL, 2015.

[22] NEVES, G. M. **Influence of the solar radiation spectrum on photovoltaic modules.** 2016. 240 f. Dissertation (Master's in Space Engineering and Technology/ Science and Technology of Materials and Sensors) - National Institute for Space Research, São José dos Campos, 2016.

[23] GUO, G. et al. **Modelling of solar photovoltaic cells and output characteristic simulation based on Simulink.** Journal of Chemical and Pharmaceutical Research. 2014.

[24] ANALOGUE DEVICES. **LTspice.** Available at: < http://www.analog.com/en/design- centre/design-tools-and-calculators/ltspice-simulator.html>. Accessed on: 12 August 2018.

[25] YOUTUBE CHANNEL: CU-ECEN4517. **3.1 PV panel Spice model and simulation.** Available at: < https://www.youtube.com/watch?v=Z2C28OXu4xA>. Accessed on: 12 August 2018.

[26] SOURCEFORGE. **QtGrace native Grace for Windows, Linux and Mac OS X based on Qt.** Available at: < https://sourceforge.net/projects/qtgrace/>. Accessed on: 12 August 2018.

[27] UNIVERSITY OF SÃO PAULO (USP). **Exponential function**. São Paulo, 2001 -

2012. Available at: < http://ecalculo.if.usp.br/ >. Accessed on: 07 July 2018.

Printed by Books on Demand GmbH, Norderstedt / Germany